FLORA OF TROPICAL EAST AFRICA

OLINIACEAE

B. VERDCOURT

Trees or shrubs. Leaves simple, opposite or ternate, at first sight apparently without stipules but vestiges are present. Flowers hermaphrodite, regular, in terminal or axillary cymes. Calyx-tube joined to the ovary and produced above it as a cylindrical tube, the margin of which is sinuate or minutely 4–5-toothed, ultimately deciduous. Petals 4–5, imbricate, linear-oblong to spathulate, inserted on the margin of the calyx-tube, pubescent at the base, alternating with 4–5 small valvate ± incurved pubescent coloured scales which close the throat in the young flower. Stamens 4–5, inserted on the calyx-tube, alternating with the petals, inserted below the scales and at first hidden by them; filaments very short; connective thickened; anthers 2-thecous, dehiscing inwardly; alternating with the stamens are 4–5 bilobed hairy swellings at the base of the petals and usually interpreted as staminodes. Ovary inferior, 3–5-locular, with (1–)2–3 pendulous ovules in each locule; placentation axile; style simple with clavate stigma. Fruit a 3–5-locular false drupe with 1 seed per cell. Seeds with spiral or convolute embryo; endosperm absent.

A single genus *Olinia*.

There are different interpretations of the floral structure and there is a good deal to be said for considering the " calyx-tube " to be a prolongation of the receptacle, the " petals " as calyx-lobes and the " scales " as the true petals. This is in fact the interpretation of Fernandes and several previous workers. Since, however, the application of specific terms to these organs has little real meaning I have retained the interpretation which a user of the Flora would probably make on first inspection. The floral anatomy has been discussed by Rao & Dahlgren (Bot. Notis. 122: 160–171 (1969)) and they agree to a large extent with the interpretation I have adopted. The affinities of the family are usually stated to be with the Lythraceae and there is indeed some resemblance in structure, but recent workers suggest that the true affinities are with the Thymelaeaceae and Penaeaceae. Discussions on this matter have been published by Fernandes (Comptes Rendues A.E.T.F.A.T. IV Réunion 1960: 283–288 (1962)), Weberling (in E.J. 82: 119–128 (1963)) and Mujica & Cutler (in K.B. 29: 93–123 (1974)). It is interesting to note that Takhtajan (Syst. Phyl. Magnoliophytorum: 304 (1966)) combines the above ideas since, although putting the Oliniaceae some little way after the Lythraceae, he places the Penaeaceae after Oliniaceae and not with the Thymelaeaceae.

OLINIA

Thunb., Arch. Bot. Roemer 2(1): 4 (1799), *nom. conserv.*

Plectronia L., Syst. Nat., ed. 12, 2: 138, 183 & Mant. Pl.: 6, 52 (1767)
Tephea Del. in Rochet d'Héricourt, Second Voy.: 340 (1846)

Characters of the family.

A small genus of probably only 5–6 species (usually estimated at about 10) in eastern and southern Africa from Ethiopia to South Africa west to E. Zaire and S. Angola. There does not seem to be any real evidence that *O. ventosa* (L.) Cufod. is truly indigenous to St. Helena, being otherwise a Cape species.

O. rochetiana *A. Juss.* in Comptes Rendues Acad. Sci., Paris 22: 812 (1846); Liben in B.J.B.B. 43: 235 (1973) & in F.A.C., Oliniaceae: 1, t. 1 (1973). Type: Ethiopia, *Rochet d'Héricourt* (P, holo., K, iso. !)*

Shrub, small tree or less often a large tree, evergreen, usually (1·2–) 4–16 m. tall but occasionally said to reach 27 m.; often a small gnarled bushy tree in exposed rocky places. Bark grey, smooth or rough, sometimes slightly peeling or flaking, often fissured; blaze white with reddish or purplish border. Branchlets reddish when very young, later pale, mostly squarish, which together with the leaf venation renders sterile twigs easily identifiable. Leaves opposite or ternate, often bright red when young; blade broadly rounded-elliptic to narrowly elliptic or rounded-rhombic, 0·6–12 cm. long, 0·5–4·5 cm. wide, rounded to acuminate at the apex, the actual apex blunt or emarginate, rounded to cuneate at the base, glabrous or minutely puberulous, paler beneath with the close reticulate venation characteristically darker or pellucid and forming a conspicuous pattern; midrib often reddish, impressed above, prominent beneath; petiole 2–9 mm. long, reddish. Inflorescences globose or pyramidal, 1·5–7·5 cm. in diameter, usually many-flowered, the branches often puberulous; pedicels very short or up to about 1 mm. long; bracts very deciduous or rarely persistent, thin, veined, ovate or oblong, cucullate, pubescent, those enveloping a triad of flowers or simple terminal flowers 1·5–6 mm. long, 1–3·5 mm. wide, larger than the secondary bracts enveloping lateral flowers of the triad which measure 1–3·5 mm. long, 1–2 mm. wide. Flowers sweetly scented. Calyx-tube and combined receptacle ± cylindrical, somewhat narrowed to the base, mostly crimson, glabrous or puberulous, 2·5–7 mm. long, minutely lobed or undulate at the apex. Petals yellowish cream, white, pink or crimson, often white at first and becoming red later, linear-oblong to distinctly obovate-spathulate, 2·3–5 mm. long, 1–2·5 mm. wide, rounded at the apex, always narrowest towards the base, mostly glabrous but usually pubescent towards the base inside. Scales crimson-pink, 0·5–1·5 mm. long, 0·5–1·25 mm. wide. Fruit pink or dull crimson, globose or ovoid, 0·5–1 cm. long and wide, usually speckled with pale lenticels, rather woody inside, marked by the circular scar remaining after the calyx-tube has fallen off. Seeds brown, subtrigonous-ovoid, 2·5–3 mm. long and wide, minutely shagreened.

UGANDA. Karamoja District: Mt. Morongole, Apr. 1960, *J. Wilson* 999!; Kigezi, *Mukasa*!; Elgon, Kaburoron, Jan. 1936, *Eggeling* 2484!

KENYA. Northern Frontier Province: Ndoto Mts., Sirwan, 1 Jan. 1959, *Newbould* 3367!; Nakuru District: Subukia Escarpment, Thomsons Falls to Lake Solai road, 21 Jan. 1959, *Bogdan* 4762!; Kiambu District: Muguga North, 18 Feb. 1962, *Verdcourt* 3283!

TANZANIA. Moshi District: Sanya sawmills, 13 Jan. 1954, *Hughes* 191!; Mpanda District: below summit of Kungwe Mt., 7 Sept. 1959, *Harley* 9550!; Rungwe District: Kyimbila, Mwakaleli, 13 Nov. 1913, *Stolz* 2288!

DISTR. **U**1–3; **K**1, 3–6; **T**2–4, 6, 7; E. Zaire, Rwanda, Ethiopia, Malawi, Zambia, Angola and Transvaal

HAB. Upland dry and moist evergreen forest, forest edges and frequently forming thickets with *Rhus* etc. in areas where forest has been degraded by fire and other human interference, finally as a relict tree in grassland; 1680–3000 m.

SYN. *Tephea aequipetala* Del. in Rochet d'Héricourt, Second Voy.: 340 (1846). Type: Ethiopia, Choa, probably near Ankober, *Rochet d'Héricourt* 18 (MPU, holo.)

[*Olinia cymosa sensu* Hiern in F.T.A. 2: 485 (1871), *non* Thunb.]

O. usambarensis Gilg in E.J. 19: 278 (1894) & in E. & P. Pf. III.6a, t. 74/A–G (1894); Engl., V.E. 3 (2): 624, t. 277/A–G (1921): Robyns, F.P.N.A. 1: 650, t. 67 (1948); T.T.C.L.: 395 (1949); I.T.U., ed. 2: 290, t. 14 (1952); Brenan in Mem. N.Y. Bot. Gard. 8: 441 (1954); K.T.S.: 350, t. 22 (1961); F.F.N.R.:

* Probably based on the same gathering as *Tephea aequipetala*.

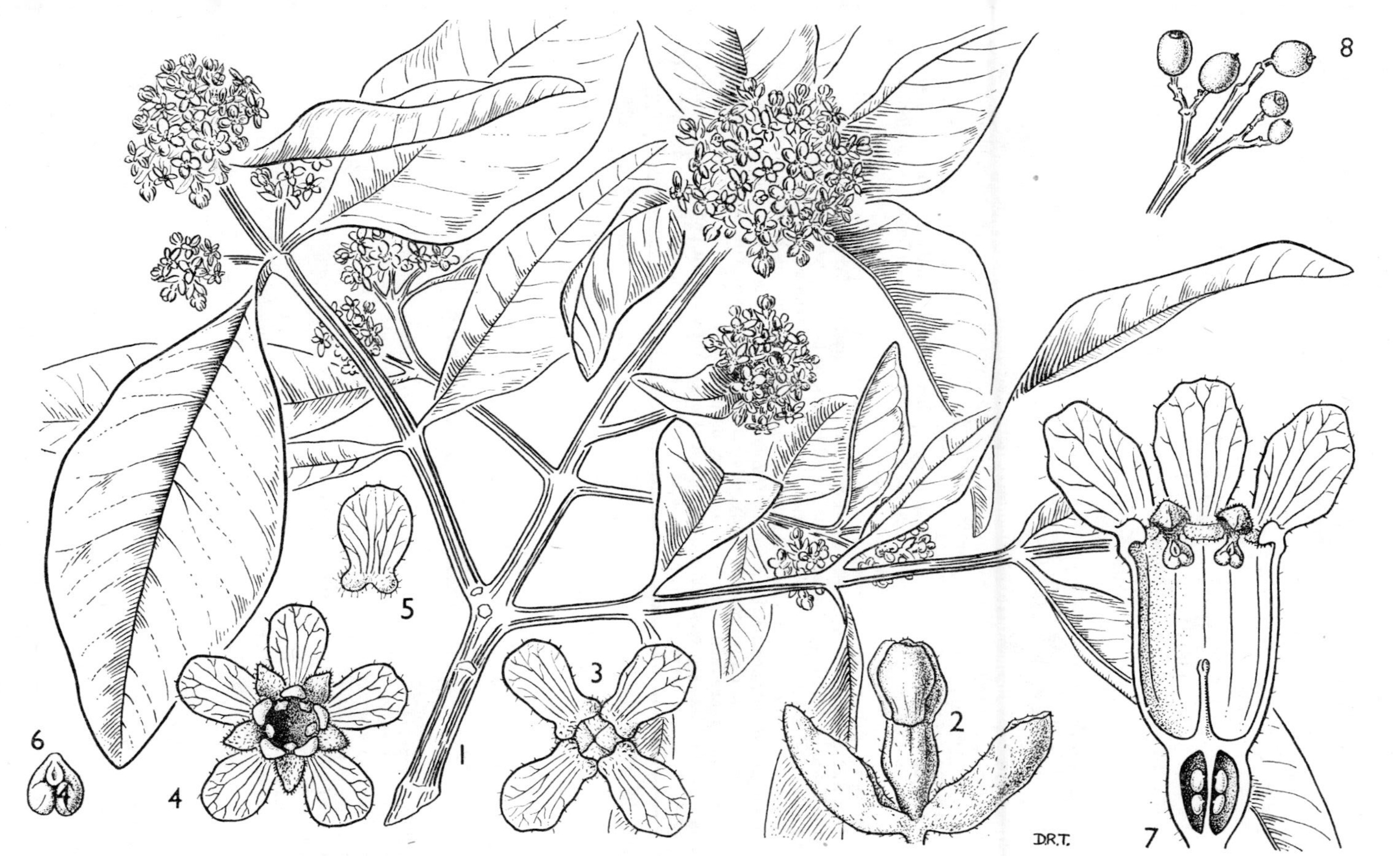

FIG. 1. *OLINIA ROCHETIANA*—**1**, flowering branch, × 1; **2**, flower bud, × 4; **3**, young flower from above, × 4; **4**, flower, × 4; **5**, petal, × 4; **6**, stamen, × 12; **7**, longitudinal section of flower, × 6; **8**, part of fruiting twig, × 1. All from *Lugard* 202. Drawn by Miss D. R. Thompson.

269, t. 47 (1962). Type: Tanzania, W. Usambara Mts., Kwa Mshusa, *Holst* 9115 (B, holo.†, BM, K, iso. !)

O. volkensii Engl., P.O.A. C: 285 (1895); T.T.C.L.: 395 (1949). Types: Tanzania, Moshi District, above Kilema, Himo R., *Volkens* 1816 (B, syn.†, BM, BR, K, isosyn. !) & above Useri, *Volkens* (B, syn.†) & Karrakia Gorge, *Volkens* 2000 (B, syn.†)

O. macrophylla Gilg in Z.A.E. 2: 575 (1913). Type: Zaire, Ruwenzori, Butahu [Butagu] Valley, *Mildbraed* 2541 (B, holo.†)

O. ruandensis Gilg in Z.A.E. 2: 575 (1913). Type: Rwanda, Rugege Forest, *Mildbraed* 1027 (B, holo.†)

O. discolor Mildbr. in N.B.G.B. 11: 669 (1932); T.T.C.L.: 395 (1949). Type: Tanzania, Njombe District, Lupembe area, Likanga stream, *Schlieben* 20 (B, holo.†, BM, BR, iso. !)

O. aequipetala (Del.) Cufod. in B.J.B.B. 29, Suppl.: 603 (1959) & in Oest. Bot. Zeitschr. 107: 106, 109 (1960)

O. huillensis A. & R. Fernandes in Mem. Junta Invest. Ultram., sér. 2, 38: 15, t. 1 (1962). Type: Angola, Huila, between Lopollo and Humpata, *Welwitsch* 991 (LISU, holo., BM, iso. !)

NOTE. Recently Fernandes and Cufodontis, in the references cited above, have suggested that up to 4 species can be recognized among the synonyms listed, but the characters used are not satisfactory and a survey of the abundant material now available suggests that one variable species is concerned. Much has been made of petal shape, whether linear-oblong, oblong, oblong-spathulate or distinctly spathulate. In the survey mentioned the shape of the petals was particularly noted and although most Ethiopian material has the sepals linear-oblong and little if any of the East African material shows this, short sepals do occur in several Ethiopian populations (e.g. *Gillett* 14824, 64 km. NE. of Addis Ababa). The populations occurring in Angola, W. Zambia and the Transvaal with short broad petals could perhaps be recognized as a variety using Fernandes' name. Some other variations not noted by the various authors may be mentioned. Trees on Elgon often have larger more persistent bracts giving a very different appearance to the plant (e.g. *Lugard* 202). In the Njombe–Iringa area of Tanzania plants on windswept rocky ridges have thick leaves with recurved margins. Until much more field work has been carried out I prefer to consider *O. rochetiana* to be a very variable species in common with so many of the trees with which it constantly occurs. The flowers and fruits are often galled, so much so, that few or no normal flowers are to be found which frequently deludes collectors into believing they have found a new taxon.

INDEX TO OLINIACEAE